UGG: The **U**ndergraduate **G**uide for **G**raduate School*

*Everything you always wanted to know about graduate school in the biomedical sciences, and were afraid to ask.

By Steve Caplan, Ph.D.

Table of Contents:

1) Introduction

2) The Scientific Career Pathway: how does it work?

3) How to build a strong CV during your career as an undergraduate

   a. Grades, especially in science-based courses

   b. Undergraduate research experience (presentations, publications, recommendations)

   c. GRE exam tests

4) Gaining acceptance to graduate programs

   a. What is the purpose of graduate school, and just as importantly, what is NOT the purpose of graduate school

   b. Gap year? Pluses and minuses

   c. Selecting a graduate program—differences between "top tier and state universities" (status and pressure), between undergraduate campuses and medical schools (stipend pay and teaching assistantships), interdisciplinary programs, names of degrees (biochemistry, pharmacology, basic sciences)

d. The Application

e. The interview—how to prepare, and how to make yourself desirable to a graduate program

5) Welcome to the University of X's Graduate Program! Now what?!

a. The single-most important career choice for a student: Finding and securing a mentor

b. Success in a graduate program—how to build a resume as a graduate student and surpassing important milestones

6. Conclusions

1)   INTRODUCTION

Congratulations! If you are reading this little manual-for-a-scientific career, this how-to-do-it booklet, then you are already a leg up on the competition! You have also chosen a very noble (Nobel—pun intended) career. But seriously, you have made a bold decision, and I am the first to applaud you. That is, I suppose, because your parents were pushing you to enroll in medical school, or law school, business school, or computer science. Something *"practical."* However, unless I dissuade you from a career path in science, you have chosen the most fascinating and meaningful career possible. It is, you say? Yes, it is. Well, at least if you can get a job! Seriously, that's what UGG is all about—giving you the best information and helpful tips so that you will thrive in science.

Several years ago I wrote a satirical article titled [“How not to get a lab job.”](#) In that piece, designed primarily for graduate students who were looking for post-doctoral positions, I tried to use real-life examples based on the types of letters and applications that I received to humorously illustrate what *not* to do in looking for a position. By all accounts the piece was a big success, and the key focus was to emphasize the importance of professionalism when applying for a position. However, surprisingly (at least to me) I also received complaints that while I was illustrating what not to do, I had failed to help those who truly wanted to learn how to better their chances of finding suitable employment.

With that criticism in mind, and after years of serving on and then chairing a departmental graduate and admissions committee, I now intend to rectify that deficit. Because those most in need of clear, concise and accurate information on graduate school and careers in science are usually undergraduates, I intend for this manual to provide a step-by-step guide to help those interested in a career in biomedical science navigate their way through the process of gaining acceptance to graduate school and onward. So, let's get started!

2) The Scientific Career Pathway: how does it work?

Students of biology are probably familiar with what has been called as the "Central Dogma:" DNA-to-RNA-to-Protein.

In analogy to that progression, for an academic career, we can say the following:

Undergraduate-to-Graduate Student-to-Post-doc-to-Assistant Professor/Faculty member

In reality, this type of progression lists a single career path: that of an academic research career. However, data suggest that less than 25% of post-doctoral fellows end up in tenure track assistant professor faculty positions when tracked for 5-6 years after graduation (Denton et al., PLoS ONE 17(2): e0263185. https://doi.org/10.1371/journal.pone.0263185). Does this mean that there are no jobs for biomedical scientists with Ph.D. degrees? No! Over the past decade, the biomedical science world has simply come to understand that an academic career path is simply one possible track within a wide range of important and very fulfilling scientific careers. Such careers include teaching at undergraduate institutions (or even high school), research careers in industry, core facility managers, work in regulation, public policy, science writing/journal editing and more. All these biomedical science careers usually require Ph.D. level scientists, and for many careers, having additional post-doctoral training is benificial. For the purpose of this manual, I will focus primarily on the Ph.D. degree, although the significance of post-doctoral studies will be addressed in part.

The typical undergraduate program in the US is 4 years, although some other countries, including the United Kingdom and Israel, tend to have more focused 3-year undergraduate programs. While there may be advantages to either system, this manual is more focused on the upcoming stages of a scientific career, and thus will avoid discussion of the differences in undergraduate programs.

What is the most important stage of the scientific career?
That is a hard question to answer, because each step is important, is an advance built upon the previous step, and for this reason, you need to do well at each stage along the way to make it to the top. For example, if you have poor grades in undergraduate studies, it'll be difficult to get into a graduate program. Having said that, usually

once you've been admitted to a graduate program, for example, your undergraduate grades no longer matter very much. But they still have some impact.

It's important to remember that once you start your career, as an undergraduate, everything you do—or don't do—can impact your career later on. Good grades make it easier to get into graduate programs. Working and impressing a mentor in undergraduate research will also help. But even once you've crossed that hurdle and been accepted to a graduate program, fellowships, assistantships, studentships and scholarships will inevitably ask for undergraduate transcripts, since this provides another way for reviewers to evaluate you as a candidate for such an award. By the time you begin post-doctoral studies, the significance of your undergraduate grades will have all but disappeared, only to have been replaced by your 'productivity' (publication record) as a graduate student. In other words, your previous track record is the most relevant, and usually has the most impact on you're the next stage in your career.

For these reasons it is difficult to define the most significant stage of your career; however, there are nonetheless two very important stages: 1) your Ph.D. 2) your post-doctoral studies. How do they differ?

Let's start with progression in an academic career: obtaining a faculty position as an assistant professor usually comes immediately following one or more post-doctoral training periods. This is the time when a scientist commits full time to research, and most faculty search committees will weigh productivity most heavily for this period of time. Why? First, because this will be the most recent research that you do when applying for a faculty position. Second, because the type of research that you carry out during your post-doc will usually be closely related to the projects that you will "take with you" or initiate in your new faculty position. Third, because post-docs are generally more independent, and the work done during your post-doc will be considered to have more of your "fingerprints" on it than your Ph.D. research, because that will be considered more "mentor-driven." Whether this is entirely accurate is another issue, but that is often how it is perceived. For these reasons, a strong post-doc in an outstanding lab, with a great and highly acclaimed mentor in a well-known institute can overcome a Ph.D. that might have been modest in productivity. The *catch* is that in order to gain acceptance to such outstanding positions, usually you need to have an outstanding Ph.D. in terms of productivity. So, as you see, it really is a cycle, with one stage dependent on the previous stage.

Graduate Programs:

Although graduate programs can include both Masters of Science (M.Sc.) and Doctor of Philosophy (Ph.D.), in the US today, the vast majority of students interested in a scientific career apply directly to Ph.D. graduate programs without first doing a M.Sc. degree. However, since publication of the initial version of this manual five years ago, there has been a resurgence of sorts in M.Sc. programs in the US. It should be noted that in many other countries, M.Sc. degrees are still considered a precursor to doing a Ph.D., and international students applying to US graduate programs often come with a M.Sc. degree in addition to their Bachelor's degree.

What is the difference between a M.Sc. degree and a Ph.D. degree, and is there any advantage in doing a M.Sc. degree first?

M.Sc. degrees in the biomedical sciences can be subdivided into two major types: 1) Thesis, and 2) Non-thesis. Since the non-thesis M.Sc. degrees do not necessarily provide a research experience, they are not especially relevant for those interested in a career in scientific research. Today there is a growing market for such programs as "feeders" for acceptance into professional programs including medical, dental and veterinary schools, etc. Accordingly, these programs are often expensive. M.Sc. degrees that include a thesis in the biomedical sciences usually involve a 2-year program of studies combined with research, with a written thesis based on the laboratory research. Unlike the non-thesis system, M.Sc. students are often paid stipends, although they can be considerably lower than the ones Ph.D. graduate students receive. In some cases, the M.Sc. graduate courses may be similar or even the same as those designed for Ph.D. students, and the M.Sc. degree turns out to be a kind of mini-Ph.D. that occurs over two years instead of five or more, so that the primary difference between a M.Sc. graduate and Ph.D. degree-holder is in the actual research experience and independence in the lab, rather than purely theoretical knowledge. It is also noteworthy that in some cases, M.Sc. students can transfer into the Ph.D.

programs at that institution within a year or so, if the student is in good-standing with grades and work ethic in the laboratory.

As noted, to obtain a Ph.D. degree in the biomedical sciences, one used to be required to first complete a M.Sc. However, over the past 20 years or so, at least within the US, the M.Sc. degree has almost disappeared from the biomedical sciences, and has been replaced by what is known as the "direct path to Ph.D.," with students moving directly from completion of their undergraduate studies to a full-blown Ph.D. program. However, depending on the country/institute in which they learned, international students often are required to first do a M.Sc. degree (either in their own country or in the US), to increase the chance of success in the Ph.D. program. While a direct Ph.D. program may significantly shorten the time until a student obtains her/his Ph.D., it's important to examine how this might affect the career-minded student in the long-run.

Let's first have a brief look at the direct Ph.D. pathway: it usually takes five years or longer to obtain a Ph.D. from the time a student enrolls in a direct Ph.D. program. Initially, the student takes 1-2 years of course work (as in the M.Sc. system), identifies a research lab by doing several months-long internships (often known as "rotations"), selects a laboratory for her/his research and spends considerable time in the lab doing research until completion of the degree (usually 4.5 years or more), becoming a bona-fide expert in her/his realm of research.

So isn't it a "no-brainer?" Why bother with a M.Sc. degree?

In many cases it *is* a no-brainer—but there are certain advantages in doing a M.Sc. degree, and there are some potential disadvantages in the direct Ph.D. pathway. First, there are some important issues to consider. For example, doing a direct Ph.D. speeds up the process of obtaining a Ph.D., and conversely, getting a M.Sc. degree slows down the time until one would obtain a Ph.D. Thus, in the immediate range, a direct Ph.D. clearly has economic advantages—one gets a better stipend and should receive the Ph.D. degree at a younger age, thus getting a "permanent job" and starting to earn a decent salary more quickly. However, it is necessary to take a step back and ask what the ***goal*** of a Ph.D. degree is. This is a key point: if you are interested in a career in science, research, academia—then the Ph.D.

on its own is rather meaningless. It is a piece of paper than some may frame and hang on the wall, where others (me included) simply leave coiled up in the cardboard tube in which I originally received it twenty-five years ago, never to see the light of day. The reason that the degree itself is limited in its importance is that to advance one's scientific career, the act of obtaining a Ph.D. has to be accompanied by a series of significant scientific achievements.

Let's return to the introduction to this manual, where I compared the "Central Dogma" of biology to the "Central Dogma" of an academic research career, and explained that after a Ph.D., an academic career oriented graduate student needs to progress to a post-doctoral position. The act of finding a suitable post-doctoral position and being accepted as a post-doctoral fellow in an excellent laboratory requires that one holds a Ph.D., but competition is stiff, and the more scientific publications, better recommendation letters, and overall achievements that an applicant has, the more likely he/she will be successful in joining a top-notch lab. This is especially true in 2023, where post-doctoral salaries continue to rise, and grant funding per laboratory has remained mostly stagnant. As a result, there are fewer positions available for post-doctoral fellows, and competition for good labs/positions is increasing.

Accordingly, imagine a scenario where the principal investigator (PI) of such a top-notch lab has a choice of either: 1) a post-doctoral fellow applicant who did four years of undergraduate studies, and a five-year direct Ph.D. program in which he published **one paper**, or 2) a post-doctoral applicant who after four years of undergraduate studies did a two-year M.Sc. program in which he/she co-authored a paper, and then spent six years getting a Ph.D., but managed to publish **two papers** and **co-authored a third paper**? With all things being equal, the advantage certainly lies with the latter applicant.

What I am trying to establish—to simplify things even more—is that the issue of "saving time" can sometimes work *against* someone with ambitions of a scientific career. In this case, the student who spent three more years until completion of his Ph.D. would (at least on paper) appear to be more qualified and more likely to land a top-notch post-doctoral slot. In many cases, researchers who can't obtain top-notch post-doctoral positions will be unlikely to later obtain faculty positions—or they may opt to do more than a single post-doctoral fellowship, with the first one designed to enhance one's reputation and publication record in order to obtain a second post-doctoral position in a top-tier lab. So, while one may think that one is saving time, this is not always the case. To be fair, I have heard a reverse argument: at least once you get a Ph.D., you get paid more—so better to do two better-paid post-doctoral fellowships than a M.Sc. degree where you earn much less. Economically, this makes some sense, but one must remember that it's also potentially riskier and harder and harder to find *any* positions as you move up the pyramid.

Another consideration with regard to choosing where or not to undertake a M.Sc. degree has to do with the selection of a Ph.D. mentor, lab and research project. I find that many undergraduates do not have significant research experience by the time they are ready to begin their Ph.D. research, and as a result, are often unsure of what they really want to study. In addition, having minimal lab experience, these students often have a harder time evaluating the research environment in the labs in which they rotate, making it more likely that despite trials in 3-4 laboratories, they may end up choosing a lab/project/mentor that is not a good fit for them. Of course, such a sub-optimal choice will greatly reduce the fun and potential success of the student in her/his graduate studies.

Masters students, alternatively, have a two-year long experience—that, even if not ideal, provides them with a very real sense of what lab-life is like, and what they appreciate and dislike in a lab and mentor. Moreover, in many cases the M.Sc. student will stay at the same institution for her/his Ph.D., and the two-year Masters program provides an outstanding opportunity to evaluate mentors and surrounding laboratories in the department and institution, allowing for better choices in the Ph.D. program.

So then, why do a direct Ph.D.? Good question! I'm glad you asked—or at least that I asked for you. First, M.Sc. programs have become more rare in the US in recent years, and in some cases students even need to pay tuition. Second, despite the potential concerns listed above, many students nonetheless do have a firm grasp of what they want to research, and for them an extra two-years feels like a waste of time. Third, despite the extra training, in scientific terms, 2-years is a short period of time to learn a lot of new techniques and advance the research in the field enough to publish a paper and rack up a documented achievement. And fourth, just like a student can choose the wrong mentor for Ph.D. research, the same thing can happen for M.Sc. research—and the net result is the same—unhappiness. Albeit for two-years rather than five-years.

Summary of M.Sc. vs direct Ph.D.

Pros	Cons
Provides research experience	*Slower route to Ph.D. and beyond*
Opportunity to evaluate mentors	*More difficult economically*
Better preparation for Ph.D. program	*Hard to publish in the course of M.Sc.*

3) How to build a strong CV during your career as an undergraduate

The academic world has a number of terms that it uses for a document detailing information on your career. For example, you may have heard the term *"resume,"* or the acronym *CV*, short for the Latin term *curriculum vitae*. Or you may have heard the term preferred by the National Institutes of Health, which is *"Biosketch."* Admittedly, there are important differences in the style and content of these types of documents; if you are interested in learning more, please go to *Science magazine*: http://www.sciencemag.org/careers/1996/10/how-write-winning-r-sum. Here I will use the term CV for collecting and collating your career progression, as it is the most basic and cumulative document that records your successes, and can easily be converted or used to make resumes, biosketches and other documents that you may need. The CV is an incredibly important and evolving document that will have a significant impact on your career as a scientist. However, your CV is also *disproportionately important* when you are an undergraduate. Why? The reason is that at this early stage of your career, there are fewer ways to evaluate you—for the purposes of acceptance to graduate school, for fellowship applications or any other awards. For this reason, the CV and documentation of your career progression and highlights are incredibly important and well worth spending time to prepare.

At this stage of your career, the precise *format* of the CV is less important; many universities actually have specific templates, formats or requirements, and I would like to focus less on the format and more on what you can do to strengthen the content of the CV during your undergraduate years. Indeed, suggestions for CV format can easily be found online, for example, here: http://www.sciencemag.org/careers/2006/10/tips-successful-cv.

Aside from the requisite information that includes your name and the basic information that goes on your CV, there are typically three areas that you as an undergraduate student can and should make an attempt to develop, if your goal is a career in science. These are: a) Grades, b) Undergraduate research experience, and c) GRE exam scores. All three of these components can be highlighted and emphasized within your CV.

a.) Grades

To be able to emphasize good grades, you obviously need to apply yourself and obtain impressive grades as an undergraduate. This does not mean that a student without a 4.0/straight A record will be denied entry to graduate programs, but it is necessary to take the issue of grades seriously. Student who achieve primarily A and high B grades will be viewed positively. A student with C and D grades will not. Yes, there are some mitigating circumstances that we will address shortly, but the bottom line is that the overall trend of a student's scholastic record is going to be an important factor in acceptance to graduate programs, and beyond. What do I mean by "beyond?" There are a variety of assistantships/scholarships that graduate students can apply for, and inevitably having a good track record academically is a crucial factor for all of these awards—although there is a slow awakening among institutions that review fellowships (such as NIH) that grades should play a major role in the review. In any event, I don't mean that a student who received a B in organic chemistry or physics is "doomed" for life, especially since it's obvious that a "B" in one college may be a harder-earned grade than an "A" in another university (or even in the same college with a different instructor). But if you receive a string of C grades, this will likely make it very difficult to obtain any type of fellowship, even if you are admitted to graduate school.

b.) Undergraduate Research

Obtaining research experience is key to acceptance to graduate programs; just as importantly, it is an excellent way for you to determine whether a career in research is to your liking. You will learn whether you enjoy lab work, independently and as part of a team. You will find out whether you have the patience and motivation to carry out long-term studies in a lab. Just as importantly, you will learn what it is like to work with a mentor, what kind of atmosphere you prefer, whether you enjoy working in a big or small lab, whether you are excited by your project and have an affinity for a certain area of research. In short, this is a very valuable experience.

Aside from your personal growth (and I'm not knocking it--that is really important), research experience is one of the factors that could potentially mitigate somewhat less-than-stellar grades. Put yourself, for a moment, in the position of a PI (Principal Investigator), the head of a lab. For the record—you may put yourself in that position for more than a moment, if you manage to launch a successful career! But for the moment: you want to accept into your lab the students who are the most capable, the most likely of succeeding at doing independent experiments and carrying out novel and meaningful research. How much would you, as the PI, weigh undergraduate grades versus research experience? Here's my take on the situation (and I have had more than fifteen graduate students in my lab): the grades are about "getting your foot in the door," whereas the research experience helps prop that door open, once you have your foot inside.

Let's take an example: If I have a student applying for our graduate program with mostly C grades and B grades, a couple A grades and perhaps a D or two—that student will have a hard time getting a foot in the door. Why? Because in an age where there is competition, graduate programs will assume that either such a student isn't dedicated enough to study hard, interested enough to do well, or will potentially have academic difficulties in the coursework. The latter is the least likely, because as we say in the field, "it isn't rocket science" and most undergraduate grades (to a degree) depend more on determination and persistence rather than genius. Next, we have a student with straight A grades, but one who has never set foot in a research lab (outside lab courses) and never picked up a pipette or designed an experiment with controls. Alternatively, we have another student who has a mix of A and B grades, but who has worked a whole summer and school year in a lab. This student has carried out a variety of experiments, presented a poster on the research she carried out, and managed to work about twenty hours a week during her senior year. Impressive, no? As a Graduate and Admissions Chair, I'd say that's very impressive, and my view would be that this student may be better equipped for success than the straight A student. Are you getting the picture?

Now that we have an idea why research experience is so important, both for personal development of the student and for admission to graduate programs, how do we document the experience to showcase? It's like the old saying: "If a tree falls in the woods and no one hears it, did it really fall?" If you worked doing research and didn't document it, will you get the credit and recognition you deserve?

The most important mechanism for documentation of your experiences and successes is your (growing) CV: every career-related event that you can think of should be documented, and if necessary, showcased within your CV. First of all, the research experience should be noted with a description of the time worked (hours per work, summer, etc.), the name of the mentor, the lab, the department, the college, institute/university. Including a title for the project within the CV, and even a brief synopsis of the work achieved would probably be a good idea, as it will show professionalism. Any presentations done should be documented, with the title of the talk, date, and forum listed. *For an undergraduate student, even a presentation in a lab meeting is valid*, as long as you accurately depict that it was a lab meeting. It shows that you have experienced what presenting your work is like in front of your mentor and colleagues in the lab. Obviously if you are fortunate enough to go to a regional, national or even international meeting, be sure to highlight your participation in these highly-regarded forums. Of course, the same is true for a poster presentation at any meeting, and even if you attend a regional or national meeting but do not present your work, your attendance is still definitely worth cataloging in your CV. If you are fortunate to have contributed significant scientific data toward a paper and you are included by your mentor as a primary or co-author on a manuscript, this is essential to include and highlight in your CV. Note: even if the manuscript has not yet been submitted, try to obtain from your mentor a tentative title for the manuscript and a list of the authors, and you can include this in your CV as a manuscript "to be submitted shortly for peer review."

Perhaps one of the most important outcomes of working in a lab as an undergraduate is having the opportunity to solicit "sponsors" or "referees" who will support your career and application to graduate programs. The primary such person is the PI of the lab that you are working in, but could also include senior people in the lab (preferably holders of a Ph.D.), such as post-doctoral fellows or adjunct faculty—or other faculty within the department with whom you interact. Again, putting yourself in the position of a Graduate

Committee Chair, who is looking at applications—if some students have strong recommendation letters from mentors who worked with them and trained them, this will be a significant advantage. Letters usually include passages about the reliability, professionalism, dedication, enthusiasm and eagerness, critical thinking, amiability, independence and general likelihood that a student will succeed in graduate studies. So, as an undergraduate researcher in a lab, there is an outstanding opportunity to impress one or more people who will be making a recommendation about your suitability for graduate school. Definitely take advantage of this opportunity; this is another potential way to mitigate undergraduate grades that are perhaps not as stellar as hoped. Strong recommendation letters from recognized scientists are about as good as it gets.

c) The Graduate Record Examination (GRE) Test

One the changes that has come rapidly since the initial version of this manual five years ago is that *GRE scores are no longer required by many graduate programs*. However, I will still address GRE scores because even five years ago, most graduate programs, for better or for worse, required students to take the GRE test and submit their scores as part of the admissions process (and some still do). I have never been a big fan of indiscriminately using these exams, because as a PI and former Graduate Committee Chair, if I find a student with solid grades in science classes, coupled with research experience and strong recommendations about motivation, reliability and scientific curiosity, I really don't think that a GRE score has anything worthwhile to add to this information. Indeed, in many cases the GRE is used more frequently to try and identify potentially talented international students. The reasons for this are that the quality and level of undergraduate courses at universities outside the US are very difficult for American Graduate Committees to evaluate. Compared to relatively similar systems in Canada, or even Europe, this is less of a problem, but it is far more difficult to assess the level and even interpret the grades of an organic chemistry class in a university in, for example, in a developing country. To further complicate matters, even research experience becomes more difficult to evaluate, because many PIs of research labs abroad in certain countries are not known internationally, and it's always safer to accept recommendations from those who one knows personally—or at least from a known entity. In addition, different cultures use a different style of recommendation letter, and within the US, PIs are more apt to feel comfortable reading and "interpreting" letters written by colleagues in the US. For example, American recommendation letters have a tendency to be understated, and go by the slogan "if you can't say something nice, then don't say it." Thus Graduate Committees will read letters looking for what is missing, rather than what is written. If it says that "the student occasionally disagrees with other students in the lab," many would interpret this as an understatement meaning that the student can't get along with others. Or if nothing is said about motivation, this might cause reviewers to wonder how highly the student is motivated.

On the other hand, letters from some countries are often effusive and glowing, making it difficult for evaluating if a given candidate is really terrific, or if that is the type of letter written for every student. And at the opposite pole, in other countries (some in Europe), PIs are particularly critical when they write such letters, being extremely direct and clear about both good and bad traits. Graduate committees evaluating applications from such countries need to take care not to dismiss students with outstanding potential, just because the standards for recommendation letters have such a high bar.

For these reasons, GRE scores became an additional factor in the evaluation process . Addressing the composition of the GRE components, how the scores work and are interpreted, goes beyond this manual, so I will simply include a link for those interested to read more about the GRE (https://www.princetonreview.com/grad/gre-information). The main point is that every institution treats GRE scores differently, but overall I would rank it like grades—necessary to get a foot in the door, but not enough by itself to get inside. But I would also, in general, rank the GRE lower than grades on my scale for evaluating (national) students.

4) Gaining acceptance to graduate programs

a) What is the purpose of graduate school?

Is graduate school for you? What is the purpose of graduate school? You can find many jokes on the internet, self-deprecating humor with graduate students serving as the butt of these jests—mostly maintaining that many graduate students choose this route because they either can't find "real" jobs or are doing so out of inertia and inability to think about their careers. But this does not make any sense; it's not only hard work to survive and thrive in graduate school—it's also not trivial to get accepted, and typically takes more than just inertia. So if you've chosen to go to graduate school, give yourself some well-deserved credit! But let's examine what the purpose is, as knowledge is power.

From a purely idealistic standpoint, graduate school will provide you with enhanced critical thinking skills, and train you to become an independent scientific researcher who can formulate and conceptualize new ideas and propose feasible research experiments to test hypotheses. While ideals are great, in today's world, you also need to think of practical things like a career that will pay the bills and keep you above the poverty line. Will graduate school do that?

In most universities and institutes, graduate students in the biomedical sciences receive a stipend that supports their living expenses—and of course pay no tuition for the educational experience. As of 2023, yearly stipends typically range from $29,000-$36,000. This amount often depends on the location of the university (in place like San Francisco, San Diego, New York and Boston, where the cost of living is especially high, stipends are usually higher, too). The amount may also depend on whether the university is at a medical center or undergraduate campus, with the former sometimes having higher stipends.

While you won't get rich on that, if you don't live in a very expensive urban locale (where sometimes students receive subsidized housing), that's not bad for someone who might be just 22 years old and just out of college. Just for comparison, when I began my post-doctoral studies at NIH 20 years ago in 1998, the typical post-doctoral fellowship was at $22,000. I felt like a king coming in with an international fellowship that paid me $26,000 at the time. Today, post-doc salaries start at about $50,000 and could be as high as $70,000 (based on institution and individual criteria), and are projected to increase further in the coming years. While there has been some inflation, obviously the increased wages for students and post-docs has outstripped the inflation and made these years of training more economically viable.

Okay—that helps for the time-in-training, or during graduate school. But what next? Is it 'worth it,' aside from the ideological mission of obtaining a Ph.D.? The short answer is that: it depends on you. If your primary goal is to earn money, become wealthy or well-off—then no. Wrong career, and you should get out into the 'real world' where there will undoubtedly be more opportunities to earn money. If your primary goal is just to "get a Ph.D. degree" because it's cool to have a diploma hanging on the wall and you want people to call you "Dr.," it's also not worth it. Ph.D. degrees have become ubiquitous ("a dime a dozen" as the saying goes) and you will not be accorded any special respect for having such a degree. That is a pipe-dream, an illusion.

In the simplest terms, the Ph.D. degree is designed to train you to think critically and carry out independent research. So if you think that you could see yourself one day running a research lab, training students of your own, strategizing and writing grant proposals to fund your research, teaching classes at a university in your field (or related to your field), publishing papers and delivering seminars—then you are definitely a good fit for graduate school.

But what if you love research, you enjoy work in the lab and the atmosphere in a university setting, but couldn't see yourself juggling all of those things that a professor running a lab has to deal with, all that direct responsibility? Is a Ph.D. still for you? My best answer is that yes, it is still for you—as long as you are aware of the caveats. There are some research positions where a Ph.D. level person is needed to help direct and carry out research in a lab. These would be ideal for the type of person described above: more research and hands-on work in the lab, less grant-writing and worrying about funding. The issue is that these types of positions are not plentiful in today's scientific world. For some reason (which I cannot fully understand), almost all funding agencies want to support people who are career-climbers, who want to move up the ladder to assistant, associate and full-professor. And yet, they do not want to support these experienced and extremely important researchers who are happy to stay in the lab doing the actual research. This is a shame, but as of now, that is how the system works. Due to the lack of these types of positions overall, and their inherent instability (continual dependence on temporary grant support), this means that should you want to go this route, you may be required to shift from lab to lab as funding ebbs and flows. Although this is true of other careers as well, it isn't the case for all careers: for example, school teachers often stay 30 or more years teaching at the same school. It should be noted that biotechnology companies and biotech industry also has the opportunity for Ph.D. level research jobs, but many of these positions are also not entirely stable and may sometimes even require relocation to maintain the job. But there are also many benefits to jobs in industry and there are opportunities for those interested in such jobs.

There is growing recognition at graduate schools that students should be better prepared and trained for science-related careers that don't necessarily include research—indeed, most graduate programs now include mandatory "individual career development plan" preparation by the students, to try and determine what career options are most suitable (based on strengths and what the student likes). One area that is particularly important is teaching: Ph.D. level researchers are in demand for teaching at undergraduate colleges and universities where cutting-edge research is not necessarily a major goal of the institution. In other words, if you enjoy teaching and interactions with enthusiastic undergraduates (which is a cycle that truly keeps many in this field happy and feeling youthful), this may be an ideal career for you. Some such positions include a research lab where undergraduate research is performed—depending on your interest and ability to collaborate with researchers at institutions with more research opportunities, you could actually end up doing first-class research even at a primarily undergraduate institution. But even if your lab is serves mainly as an opportunity to introduce eager young students to the beauty of scientific research, you may derive great satisfaction from helping to train the next generation of great scientists.

b)   Gap year? Pluses and minuses

Today it is common for many undergraduate students to do a "gap year" before proceeding to graduate school, or for that matter, medical school, law school or any other professional school. Is that a good idea for those potentially interested in a career in science? In my opinion, there are quite a few advantages to a gap year, with relatively few disadvantages. I did a gap year in the early 1990s and backpacked through South America. Aside from hiking the Andes, visiting the Galapagos Islands and seeing a new continent, I used the time to gain some perspective on which scientific area I wanted to work in. Having finished my degree in biology, I knew that I was more interested in biochemistry, microbiology and genetics than physiology, zoology, ecology and botany. But I had no clue how to narrow my focus down and choose a select area. Not only did scaling mountains help me decide what research I was most interest in, but coincidentally I met a fellow traveller whose father ended up being my co-adviser for my Master's thesis. Small world!

However, even if you are not the adventurous type, and even if you don't befriend someone whose father becomes your thesis adviser, gap years may still be useful and appealing to you. Many gap year students use the time to find work as a "technician" in a lab—basically getting paid (albeit not exorbitantly) to work and attain research skills in a lab. This has some potentially terrific advantages. First, it is an outstanding opportunity to see if research is really for you. Do you enjoy the challenge? Do you have the patience and tenacity to push your project forward. Do you get up in the morning excited to go to work? Or is every work-week a tiresome exercise where you wait to reach the weekend? The answer to these questions should provide important clues as to whether you will enjoy a research career. Second, it is an outstanding opportunity to see what kind of laboratory and mentor are a good fit for you. You will see whether a small, focused lab with a younger mentor who is more 'hands-on' is to your liking. Or whether a large lab, where the mentor has assigned a senior student or post-doc to work with you is a better fit. You will also see what type of temperament of a PI you are comfortable with: a PI who is easy-going and laid back, or one who is constantly engaging you and pushing you to achieve more results.

In addition to helping you to learn more about your likes and dislikes, such a gap year in a lab also provides an outstanding opportunity to further advance your goal of being accepted to a (specific) graduate program. Typically you will be able to significantly add techniques that you master to your growing CV. Hopefully, you might also complete a research project that will be published, giving you an additional accomplishment to add to your CV. You would definitely have an opportunity to solicit positive recommendation letters from your PI and possibly other faculty members in the department where you work. You would even have time to practice and retake the GRE exams, *if necessary*. In short, the gap year is full of pluses for an aspiring graduate student.

But are there any disadvantages to a gap year? Not really. Yes, you will be lengthening the route to your ultimate career job by a year—but really, in the scheme of things—and in a career where after a Ph.D. scientists are doing not one, but often 2-3 postdoctoral stints before (hopefully) landing a good position, what's another year? In my view, it's negligible, especially when weighed against the potential benefits. Is a gap year sometimes superfluous? The answer is that for students who have considerable lab experience from their undergraduate studies, and who already have a good sense of their scientific interests and are more certain of their career goals, a gap year may not be as helpful.

c) Selecting a graduate program

When you have convinced yourself that you really do want to obtain a Ph.D. in biomedical research, the time has come for you to consider *which* graduate program(s) you would like to join. As always, with such an important decision to make, it is wise to remember what the goal of graduate school is. If we return to the previous sections of this manual, we find that for a successful academic career, your Ph.D., is one of the most crucial experiences. Aside from learning critical thinking, training in a variety of methodologies, gaining experience in a lab and much more, this period of time is required to launch you into a top-notch post-doctoral position when you are done. In fact, your post-doctoral position will have a far more significant bearing on whether you ultimately obtain the "Holy Grail," a faculty position, than your Ph.D.

Obtaining an academic position often rides on having a string of high impact publications and outstanding recommendation letters from internationally renowned researchers. However, obtaining an academic position also means that you have to get accepted into a really good lab, and in many cases it can help if that lab is situated in an institution that has a critical mass of outstanding researchers. On the other hand, for a Ph.D. degree, there is an argument to be made that the overall institution and environment (critical mass of outstanding labs) has less of a bearing, as the individual lab/mentor has a much greater role. For many students, a graduate program at a state university (for example, a state university) may be just as good, or even better, than a graduate program an a more highly-ranked institution.

The overall guiding principle is that for a Ph.D., it is the individual laboratory (and mentor) and your success in that lab that is crucial—

not the name or ranking of the specific institution. So if a given university, whether it is a state or elite institution, does not have a lot of good, high quality research labs (and mentors) that are available to incoming students, then I probably would not recommend it. One might argue that elite institutions have more high quality research, and in some cases, that might be true. But most state universities also have robust and well-funded research missions, and a plethora of good mentors.

With me so far? Good. But then you might logically ask: "Okay, but if I can choose either an elite or state university with good research and good mentors, why wouldn't I automatically just choose the more elite institution?" And that is an excellent question!

My answer is that you might choose the elite institution, and that might be an excellent choice—but for many students the state institution might be just as good, with several additional potential benefits. First, in many cases at the more elite institutions (and some state institutions), the institute charges the PI tuition when they take a graduate student. Yes, you get your stipend, and no, YOU DON'T pay the tuition yourself—the lab that accepts you does. Does that matter to you? Perhaps it might: since tuition can be very expensive for the PI, essentially a student in the lab might 'cost' the PI more than employing a post-doc. Accordingly, there may be some PIs who are faced with a choice of accepting a student or a post-doctoral fellow. Given the learning curve of students who are just starting out, and their intense course work in the first year or two, some PIs may prefer an already-trained post-doc who might be able to advance the project more rapidly. This, of course, could decrease the potential number of available labs opportunities at such institutions for students. Having noted this, strong institutions often have more "training grants" available—these are grants that cover the stipend of a number of students, and also provide a training structure to the student. Nonetheless, this may leave prospective students with fewer opportunities to find good labs/mentors than originally anticipated at the elite institution. In addition, this often leaves the laboratories at such institutions weighted more heavily toward an environment conducive to post-docs, rather than Ph.D. students. For example, students typically need more hands-on training than post-docs, but in a lab with 6 post-docs and 1 student, the PI may treat everyone more like a post-doc, with higher expectations. For highly independent students, this may be perfect, but for some students this might be a harder environment in which to succeed, at least initially when more training is required to get started.

In addition to potential differences in opportunities, some smaller institutions might have additional advantages compared to their more prestigious counterparts. If you are the type of person who thrives under high pressure, perhaps this will not be an advantage to you, but there is typically a somewhat higher degree of pressure at the larger and more elite institutions. For example, almost all American universities have a "qualifying exam" (sometimes known as a comprehensive exam) for students usually between their first to third years, as a mechanism to ensure that students are deserving of being 'candidates for the Ph.D. degree.' These exams usually take the form of writing a type of grant proposal, and later defending the proposal on both conceptual and experimental grounds in front of a faculty-based examination committee. In most state universities, attrition is generally low, and often self-induced (the student realizes that this is not really what he/she had thought it would be). However, some institutions the failure rate may be considerably higher, increasing the pressure on graduate students. Obviously, if you are a bright and determined student who thrives under pressure, this will not be a drawback for you. However, there are many very talented and capable students (who will go on to do great things in scientific careers) who feel that they would rather reserve the pressure and stress for their own scientific research projects and their publications, and have no desire to include any additional (and perhaps unnecessary) stress into their careers.

A second general consideration regarding the choice of graduate programs, relates to the *type* of institution, rather than whether it is considered an exclusive or prestigious one. For example, the atmosphere at universities with undergraduate campuses differs quite dramatically from those that are exclusively graduate institutions or medical centers. Confused? Allow me to explain. If you are reading this manual, then you will likely be familiar with colleges and universities where undergraduates study. Often (but certainly not always) these institutions are located in small, pastoral college towns. They frequently have "quads" or "malls" or large central grassy areas where students can enjoy outdoor activities in nice weather. While not exactly a 'carnival atmosphere,' there is certainly a somewhat relaxed, calm feeling. I have seen and heard of professors bringing their dogs to work or to lectures, and students playing with the dog during breaks. Bicycles are common on campus. Faculty often come to work in the labs in t-shirts and casual clothing. And the presence of undergraduates on campus, and potentially in the research labs can be invigorating.

Medical centers (note: when isolated from the undergraduate campus), where medical schools are based, are often found in larger towns or cities, and not infrequently, for historical reasons, often in less attractive urban areas. These centers are usually bustling, with hurrying professionals scurrying about, doctors in scrubs, administrators in suits, nurses, patients, etc. At medical centers faculty typically dress more formally; that isn't to say that all PIs, post-docs and students walk about with suit and tie, but overall the atmosphere is deemed as 'more professional' than campuses with large undergraduate student bodies. In addition, due to the more urban localizations, grassy outdoor areas and walking paths are often limited or non-existent, with concrete and high-rise research buildings taking their place.

You might be thinking: "Well, those are just superficial differences. It doesn't matter to me." And you may be right. But there are additional differences that might influence your Ph.D., and it's good to be aware of them. For example, many of the undergraduate campuses are in dire need of graduate student teaching assistants to do recitations for undergraduate courses, or run lab-based courses. Is this good or bad? That depends on you: some graduate students are extremely excited about the opportunity to gain experience teaching. They may be interested in a career at a primarily undergraduate institution, which is heavy on teaching and has minimal research. In other cases, a graduate student may still be interested in a research career, but feels that gaining teaching experience will present some advantage down the road in obtaining a faculty position. Every institution differs in its requirements, but it should be noted that many graduate programs at campuses with undergraduates require students to be teaching assistants for one or more semesters, as part of their duties and in order to receive their stipends. Is it ever bad to have to teach? The only potential disadvantage is that the teaching can take up considerable time, and slow your research down. If your work is highly competitive, it could allow your competition the opportunity to 'scoop' you, so a heavy teaching burden is not always helpful. It could also take longer to obtain your Ph.D., or longer to publish papers, thus making it more difficult to find a top-notch post-doc position. From an economic standpoint, a longer Ph.D. can also mean more time until a more permanent job and better earnings.

Medical centers, on the other hand, often have more limited opportunities for graduate students to experience teaching. While this frees up more time for research, and there is not a requirement for teaching to receive your stipend, in some cases students can't find teaching hours and obtain the experience that they desire in a classroom. The differences between undergraduate campuses and medical schools noted here are not meant to persuade or dissuade graduate students that either type of campus is "better," but rather merely to point out some of the key differences, in case they help you to make a decision.

Many or perhaps even most American institutions now have graduate programs in the biomedical sciences that are "integrated programs." Mostly, what that means is that instead of a department-based program (i.e., "Biochemistry," or "Physiology") they offer a program that essentially allows the student to rotate in laboratories that belong to different departments. The primary advantage of these integrated programs (sometimes known as "umbrella programs") is that if a student is unsure of his/her precise research interests, then they are not locked into the research that is done only in a select department. In addition, even if there is a wide range of different research carried out in a given department, having more choices gives a greater chance for successfully finding a good lab (especially when not every lab may have funding to accept new students in a given year). In addition, since research has become so interdisciplinary, students should understand that often the departmental (or program) names really have little meaning. For example, one can be a cancer biologist in a biochemistry department, a biochemist in a pharmaceutical sciences department, or an immunologist in a cell biology department. And so on. Thus being in an integrated program allows maximal flexibility for incoming students.

d) The Application

Eligibility:

*A four-year US B.Sc. degree or equivalent from an international institution is required.

*Many institutions have a minimal GPA (3.0 out of 4 or 3.5 out of 4, for example), but obviously the higher the better.

*Some (but fewer and fewer) require GRE scores.

*International students usually are required to demonstrate proficiency in English (TOEFL scores or equivalent).

Application Submission requirements:

*Some institutions/programs charge fees for applications, whereas others do not. Fees can range from around $25-$100.

*Official transcripts. Note that many US institutions require that transcripts be "evaluated" by a vendor/company that provides information on the content and level of academic courses at that institution.

*Recommendation letters. Most programs request about three letters of recommendation for an applicant.

*Curriculum vitae (CV), listing pertinent information on career to date, achievements, awards, and experience.

*Personal narrative. Different programs define these differently, but a common theme is explaining why one would like to do a Ph.D., and highlighting how the student is motivated and equipped to deal with the challenges of a Ph.D. program.

Important:

Graduate programs will be impressed by mature student who do their homework before applying. Accordingly, a well written personal narrative should include reference to several faculty members whose research is of interest to the applicant. Here is a brief example of such a reference:

"Having worked as an undergraduate in fundamental research addressing how mitochondria undergo fission and fusion, I am particularly interested in the research of Dr. XXX in the Dept. of Biochemistry, whose recent studies on apoptosis and mitophagy are a natural extension of my own interests, and I would be excited to

discuss opportunities to develop these and other avenues of research further, with Dr. XXX and her laboratory."

In addition, highlighting specific reasons why a given department of program is a good fit for you, the applicant, can also be helpful.

How many applications should you submit?

There is no limit on the number of applications one can submit, and overall, the more applications submitted, the more likely there will be a greater number of options to select from. However, since some programs/institutions do charge application fees, and since (as noted above) a personal narrative has to be specifically tailored for each and every program, relating to faculty at that institution, submitting a large number of applications can be both expensive and time consuming. Based on experience, I think submission of 10-20 applications probably casts a wide enough net without being overwhelming for the applicant. One consideration is that different institutions have different criteria—typically the more elite institutions will have greater competition and be harder to gain acceptance. Accordingly, I would recommend that applicants submit to a variety of different institutions, with different "rankings," to increase the likelihood of getting interviews (and accepted) to at least some programs.

e) The interview

Congratulations! You applied for a graduate program and have been selected for an interview by a graduate program. How do you prepare?!

The first and most important thing to remember is that generally, they really want you to join and succeed. After all, they are going to be spending money for you to drive/fly in, and they will put you up at a hotel and feed you (maybe not wine and dine, but still, they are shelling out money for this interview). The graduate committee that is "evaluating" the students that are being interviewed is well aware that students typically apply to a cluster of different graduate schools and programs—and that if they like and chose you, then most likely other programs have also offered you an interview. And in the capitalist system, you—the student—are in demand. What this means is that just as you want to be on your best behavior and impress the graduate committee, they too want to impress you that they have a robust graduate program. In other words, it is a two-way street.

Having said that, there may be more competition to gain acceptance to certain graduate programs, and to ensure that you are accepted to the maximal number of places and given as many options as possible, you should not take the interview process lightly, but you should go well prepared. So what is the graduate (and admissions) committee going to be looking for?

That should be fairly easy to estimate—and of course, knowing what the committee is looking for should make it easier to prepare. The committee already has your undergraduate grades, recommendation letters, lab experience, personal narrative, etc. Since they invited you for the interview, they must have liked what they saw. Now is their opportunity to validate, as best as possible, what they have read. And to look for any signs that you might not be ready for graduate school.

One of the easiest and most important things to prepare for is explaining what you have worked on in your undergraduate research. The committee will want to see primarily that you can rationally explain *why* you worked on what you did, why it is a significant problem to address—and of course a little bit about what you actually achieved (or at least tried to). The most important thing is to be able to explain the rationale for your research project and how you set about addressing the question. Now—if you would like to prepare really well for your interview, here is a tip: you will have received an itinerary or schedule, and you will know in advance who you will be meeting. Sometimes it will be the whole committee together (maybe 4-5 professors) which can feel somewhat intimidating. Often it will be a series on one-on-one meetings with committee members or faculty in the department or program. In any case, being able to somehow relate your research to the research done by individuals in that department will go a long way toward convincing them of your suitability for the program. So for example, when explaining your undergraduate research, if you are able to say (for example), "Our goal was to determine how protein X is expressed in cells treated with Y, and to see if this induced their aberrant growth...." and then perhaps work in a statement "I realize that it might have been interesting to test protein Z in this system, which I know Dr. AAA is working on, and that might be interesting because these two proteins could potentially synergize because..." Showing that you are diligent, did your research and took the time to look up the research done by faculty in the program, and that you can integrate and connect things in a conceptual manner, are things that are likely going to make a very strong impression on the graduate committee.

What else is the committee looking for? They are looking for someone who will fit in, who isn't a 'trouble-maker,' who will get along with fellow students, faculty and the department at large. If you possess reasonable social skills, you will have an easy time satisfying any concerns simply by being pleasant, amiable, curious and good-natured. If you are like me (and a number of scientists) and being social isn't natural for you—then it would be a good idea to work at it in advance. It is important to have a firm handshake (or elbow bump, in the Covid era), to look people in the eye, to smile and show that you fit in and are comfortable in your own skin and around others. Being respectful is good, but being too deferential (for example, like you are in the army and saying yessir-nosir in monosyllable tones) may raise flags that you might be too shy, quiet, reserved and have trouble fitting in and acclimatizing socially. So it's important to strike a balance of respect without being overly deferential.

How should you dress? As someone who prefers very casual attire at work, I may not be the best resource for this section. But in brief: you don't necessarily need a three-piece suit or fancy dress, but you want to give the impression that you take the interview process seriously, and that you are neat, tidy and organized. As a graduate committee chair, I would probably not be overly impressed by fancy clothes, but would be more comfortable interviewing prospective students wearing clean, neat clothes without rips or holes. The goal is not to have your clothes stand out or make a statement, but rather to blend in.

Some institutions invite an entire group of potential students for an interview at the same time. This usually includes an overnight stay near campus, lunches and dinners with the admissions committee members, Ph.D. students in the program and other interested faculty. In some cases social events are arranged, with trips or hikes or visits to local attractions. Students who attend such interviews should be aware that despite the fun and relaxed atmosphere, students from the program may be asked their impressions of the interviewees, so they should maintain a professional behavior and not be tempted to talk about other interviews and/or offers.

What are some other likely questions that you should be ready to answer?

1) What are your career goals? Be ready with a good explanation of your future plans. It's perfectly legitimate to say "I'm taking things one step at a time," but do be ready to say what possibilities lie ahead for you. If you just say "I want a Ph.D. so people will call me 'doctor,'" that won't be very helpful. If you explain that you may want to teach at an undergraduate institution, or do a post-doc and then either go to academia or industry, that will be more helpful.

2) Why did you choose this graduate program? This is your opportunity to explain what your interests are, scientifically. Answer such as "I like the weather in this area of the country," or "It's close

to the beach" are not a good idea, even if they are partially true. The best answers will come with you providing some detail about specific laboratories and why the research in those labs is of interest to you. It is okay to mention that you have friends or family in the area, or a significant-other who was accepted at the same institute, *as long as this is an addendum to your answer and not the main reason.*

3) One question that I have frequently asked students who apply to our graduate program is the following: "How will you measure a successful Ph.D. after five years or so, if you are accepted into our program." The reason I ask this question is to ascertain that students really understand what they are signing up for. A surprising number of students really have no idea what constitutes a successful Ph.D. They either haven't thought about it, or heard enough about graduate school to be able to answer this. A number of students mention 'getting their diploma' as being successful. Others talk about learning techniques. A more mature applicant will say the following: "I will have learned to think critically and carry out independent research, and this will be validated by my having published first-author papers and made a name as someone who has contributed to

the field." It is very important to show the graduate committee that you fully understand that the currency of success in graduate school is publications, along with presentations at meetings and recognition—which will also lead to a mentor who will provide you with a strong recommendation letter.

Following the interview, it is good form to write a polite email thanking the graduate committee or chair of the committee for the opportunity to meet with them and learn more about the program first-hand. Highlighting specific meetings is encouraged. For example: *"I was grateful to have an opportunity to meet with Dr. AAA in person and hear more about her exciting studies and opportunities for projects on the role of mitochondrial membrane proteins in the regulation of apoptosis."* It is also advisable to write to the individual professor(s) with whom you met and were especially impressed, because they may also have sway in the final decisions of the graduate committee.

Overall, most interviews usually lead to offers for applicants, and with multiple offers, then come the hard part—making a decision! But those are good troubles to have…

5. Welcome to your new graduate program! Now what?

As discussed earlier, in the scientific career, success in the Ph.D. program is a crucial pre-requisite for healthy progression. Since the Ph.D. student is very much dependent on the mentor and the laboratory for guidance throughout the Ph.D., this means that the success of a student is, to a large degree, reliant on the success of the mentor. If a good, motivated student enters a good laboratory with a positive mentor, this almost always leads to a successful Ph.D. On the other hand, even students with outstanding potential are unlikely to be successful in a poor laboratory environment. *No matter how talented, motivated, and bright a student may be, without scientific direction and training, and without support and the mentor pushing to publish, a student will unlikely be able to achieve much that will help advance a scientific career.* I have seen a number of cases where an unfortunate choice by a student has led to stagnation and the end of what might have been a productive career in science. What this all means and boils down to is the following: *you must make every effort to choose a good mentor, and this first decision has vast influence on your future.*

a.) The single-most important career choice for a student: How to find a good mentor

As mentioned earlier, most graduate programs (at least in the US) are designed for advancing directly toward a Ph.D. without the requirement of obtaining a Master's degree. And in the majority of cases, students join a department or program, and within the first year or so, are required to set up a number of short trials in host laboratories, usually known as rotations. These rotations typically last anywhere between six weeks to several months, depending on the individual program, and they provide an excellent experience and window for the prospective student to determine how suitable the mentor (and the lab) is. However, one important point that is too often ignored by students, is that *the rotation period also serves as an opportunity for the mentor to observe the student*, and in turn, decide upon his/her suitability. In other words, prospective graduate students, beware! Rotations are a two-way street, and you had better look both ways before crossing!

Finding rotations:

Since the key to finding a suitable permanent mentor for the Ph.D. depends on first finding suitable rotation options, the identification of good labs in which to rotate is a crucial step in the process and cannot be taken lightly. Even students who have homed in on a very specific laboratory that they would like to join and have set up a rotation in that lab should not discount the significance of the other two or three rotations that they must do, or 'waste' them on laboratories that are suboptimal options. Why? For several reasons: the laboratory that the student initially desires may fill up and have no available space by the time the student completes his/her rotation, or alternatively, the student may realize that the atmosphere in the lab is not as good as he/she originally thought. Or, the mentor may not be satisfied with the prospective student (even if the student is happy with the mentor). Or a funding issue might arise and the mentor realizes that he/she can't accept a student at that time. In any case, one never knows the outcome of a rotation in advance, and having reasonable backup opportunities and alternatives is a wise course. So the optimal situation is to identify three or four rotations, all with potential—meaning that at least at the outset, each laboratory rotation might lead to selection of a mentor and lab.

What should a student look for in searching for a rotation?

Many students at this early career stage do not yet have a very good overview of the different types and wide variety of biomedical research opportunities available. While a small minority may have decided on a specific disease-related focus (for example, they prefer to research a type of cancer or heart disease), often for personal/family reasons, when looking for a rotation many students are fairly open to experiencing different research avenues. One important point is that *placing too much emphasis on specific scientific research areas in the search for a Ph.D. laboratory and mentor can lead to suboptimal selection.*

Why is it an error for the student to focus primarily on the research project in selecting a mentor? For several reasons: 1) As noted, students often lack a general overview of different scientific research areas to really know in advance if they will like the research or not. I would concur that a student who is not particularly interested in physics, chemistry and math will be unlikely to thrive in a structural biology lab, for example. And obviously a student who knows he/she cannot work with animals should stay away from labs that exclusively do animal studies. But in many other instances, students can find that they like (or dislike) projects once they begin to work on them. Excitement often comes with greater knowledge of the field, and as students gain independence in their research, they often become enamored with their selected area of study. However, in contrast, if the atmosphere in the laboratory (or with the mentor) is toxic, usually no matter how interested the student might be in the field, the Ph.D. is unlikely to end well. 2) A Ph.D. project is not likely going to be the area of research that those pursuing academic careers will continue with once they have completed their Ph.D. dissertations. Most new faculty who come to a university as assistant professors to start up their own laboratories will bring projects that they begun (and often squirreled away) during the course of their post-doctoral training, not from their Ph.D. training. What is the bottom line? *I am suggesting that a student is likely to be happy and succeed in a lab with a good mentor, even if the initial project appears less exciting.* This is a key point to consider, and when choosing a mentor, it is important to reflect on the goal of Ph.D. training.

What is the goal of Ph.D. training?

I know we discussed some of these points earlier, but certain things are worth emphasizing. As an idealist, one might say that the goal of Ph.D. training is "to gain a greater understanding of the world around us," and so on. But let's face it, altruism is hardly at the top of most students' bucket list of reasons for doing a PhD. Given that the primary reason for enrolling in a Ph.D. program—aside from personal interest, critical thinking skills, communication skills, and everything that comes with learning to be an independent scientist—is usually to kick-start a career in academia or industry, this leaves us with the next important question: What is *the measure* of a successful PhD?

And why all this verbiage about the goal and measure of success of a Ph.D. when I am supposed to be providing guidance for how to find and obtain a rotation and Ph.D. mentor? It's simple: students need to find mentors who are equipped to provide them with the ability to succeed—otherwise that Ph.D. diploma ends up being a useless scrap of paper. And—not all Ph.D. mentors are equal in that respect. Not even close. So what is the coin of success and how do students identify it in a prospective mentor?

How do we measure a successful PhD?

As I mentioned in the section of 'interviewing for a graduate program' above, *how we measure a successful PhD* is a question that I asked nearly every applicant to our graduate and admissions program when I was the chairperson. Very few students at that stage were able to communicate an answer that I viewed as realistic. Many would say, "I will have learned a lot of techniques." Others would say, "I will be able to run my own lab." Perhaps the latter answer is a step closer to what I was hoping to hear. I tried to simplify things and ask: "How will a future post-doctoral mentor rate your Ph.D. as successful, if he/she is considering hiring you among other applicants?" Some students would get a little closer to the idea and say "based on my recommendation letters." Still, very few hit the nail on the head—and this is one reason that I decided to put together this little guide—so that students would be more aware of the real-world expectations of graduate school. The answer, of course, (although there are many versions of this) drills down primarily to one thing: your *productivity*, largely assessed by looking at your *publication record*.

For simplicity sake, I will avoid getting drawn into difficult arguments as to what constitutes high productivity (there are different philosophies over the relative significance of the actual number of papers published, or so-called impact factors and prestige of the journals in which they are published). Suffice to say that if a student completes his/her PhD with several first-author papers that are published in respected peer-reviewed journals, this is probably the most important stepping-stone to open the doors to top labs for post-doctoral studies. *For a student who aspires to an academic career as an independent scientist, acceptance to a top-tier post-doctoral laboratory is perhaps the most important step.* Accordingly, this means that for such an aspiring student, *the practical goal of a PhD is to position oneself for acceptance to the very top post-doctoral positions.*

In a sense, then, the optimal Ph.D. mentor and lab will provide an opportunity for a student to learn, grow, and mature scientifically, develop critical thinking, lab skills and techniques, communication skills (oral and written), but just as importantly, provide an opportunity for the student to excel by publishing first-author research papers in peer-reviewed journals. Published papers are the 'currency' of science, and in all likelihood, a student who publishes 3-4 solid first-author papers will have doors open to almost any post-doctoral laboratory. A student who is unproductive will have more limited options, and those options that remain will likely be with mentors/labs that are not as likely to lead to major career advancement. Students who do not excel or at least perform well in their Ph.D. research are rarely able to "catch-up" and become independent scientists.

So how does one find the right mentor?

It is incumbent on the student to do due diligence and homework to identify such mentors. No mentor will say to a student, "Don't come to me, my students rarely publish—" it is up to the prospective students who are looking for rotation options to do *research* (remember that word!). How? There are a number of ways: the most simple way is to search the [PubMed](#) at the National Library of Medicine. By entering the mentor's full first and last names, the student can get a readout of all papers published since about 2002. That is long enough—primarily the student should see what has been published in the past 7-10 years. (Note—if the mentor has a very common name, the student will have to search each manuscript and examine the author affiliations to see if the university fits for that mentor, otherwise that paper might be from a different person with the same name). This requires an active effort by you, as evaluating a mentor's productivity in this manner needs careful research.

Is my prospective mentor productive and publishing well?

First, just seeing that the mentor has published ~100 papers in this period is not a sufficient measure of productivity. Papers that truly come from the mentor's lab have the mentor as the *senior author*, usually the last author on the list, and typically note that the mentor is the *corresponding author*. If the mentor's name is placed anywhere else within the paper, or he/she is not the corresponding author, it is probably not really relevant for a prospective student, as it means that the work was a collaboration and (some/much of) the actual research was done in another lab. Next, the student should make sure that most of the papers are research papers (rather than reviews of the literature). Reviews indicate the mentor is widely known and respected in his/her field, which is a good thing. But if 90% of the papers published in the lab are *just* reviews, this might be worrisome.

Another thing that a student can and should do is try to determine what current students and recent graduates in the lab have published. Most graduate programs have lists of students assigned to individual labs, and many individual labs have websites and/or have lists of current and former students. My suggestion is to take those lists and very carefully use the PubMed to determine how well these students have published with the prospective mentor. Usually a pattern emerges: in a lab where strong motivated students do extremely well, even average students tend to do well, and most students in that lab are likely to be quite successful. Beware of labs where students publish infrequently.

Another way to evaluate a successful mentor/lab is to see where former students end up. If they have mostly left science, this does not bode well. If the mentor has had the lab long enough, there should be a trail of students who have gone on to excellent post-doctoral positions (searchable on the web with some diligence) and even on to faculty positions, or alternatively, good biotech/industry positions, or other interesting jobs within the scientific community. Often this is listed on lab websites, but can be checked by searching online.

Information from current lab members and from other students in the graduate program and institute can also be helpful in filling in the blanks, but it is usually best to complement this information with your own online searches. Students in the lab may be cagey or unwilling to talk freely about their mentor (some may not want to feel that they are being disloyal). Former students, even if you catch them on the phone, are still dependent on recommendation letters and may be similarly reserved about frank assessments of the mentor.

What about funding? Should I ask a mentor if he/she has funding?

Typically a mentor will need to be able to pay for a student's stipend for the duration of his/her Ph.D. studies, and of course be able to afford chemicals, biological reagents and equipment for the research. However, this is usually not an issue—most departments or programs vet their mentor pools and will only allow mentors with stable funding (or at least a history of stable funding) to recruit a student. So it is certainly worthwhile asking, but usually a program or department will assume responsibility if the mentor runs into trouble. Even in cases where the mentor decides to move to a different institution, procedures are usually in place to find a solution for the student, if he/she does not want to move with the mentor. Generally, if a laboratory has been productive over recent years, chances are that funding is less of an issue.

What about the differences between large labs and small labs?

When looking for a post-doctoral position that will hopefully later lead to a faculty position, most of the top-tier laboratories will be medium-sized to large. Very few will be small. However, for graduate students, where we are concerned with strong mentorship and a different set of goals, small laboratories can also be outstanding places for a Ph.D., with potential for much one-on-one mentorship. The key is in the mentorship, and the same criteria apply (productivity, the mentor's overall commitment to the advancement of his/her students, etc.). In larger laboratories (more that 8-9 people), it would be important for the student to make sure that there are delegated senior people in the lab (post-docs or senior students) to help out in the day-to-day mentorship, because the head of a lab with 20 people will not have the time to discuss research and progress with a new student on a daily basis.

What about new or younger investigators?

New or young investigators can be an excellent choice for a Ph.D. student (in fact, I chose a new investigator and it turned out very well for me). They are usually very ambitious, full of energy and drive, and often spend time more in the lab in the early years training personnel and even doing some bench-work before succumbing to administrative duties later on. The only issue is that it is harder to "vet" new investigators, as they do not yet have a track record handling students or even in many cases, publishing as senior authors from their own laboratories. However, the innovative nature of their research and their motivation to succeed usually makes up for the lack of experience and track record.

I have identified a potential mentor who is productive and has successful students—how do I obtain a rotation?

The most important thing in setting up a rotation—and I can't emphasize this enough—is to be professional! A student who is unprofessional and lazy in approaching a mentor risks losing that rotation. Remember: there may well be pressure on the better mentors, and if it is a *"mentor's market"* –meaning the mentor has multiple students who are interested (for what might be a single

Ph.D. position)—then students who make a weak first impression might not even be offered a rotation. Again—as students evaluate mentors, mentors are also evaluating students.

Usually the best way to first approach a mentor is with a grammatically-correct and properly-spelled, well-written email query. It is particularly important to spell the mentor's name correctly. Put yourself in the mentor's shoes—if the student who wants to do a Ph.D. in your lab can't even get your name right, how will he/she be able to get the right chemicals in the right test tubes for the experiments to work? You may be in a hurry, or think that it isn't important, but believe me—as a mentor—these are crucial points.

The email may be brief, but should ask about a meeting to discuss a potential rotation. While it isn't necessary to go into minute details about why you are interested in the lab/field, it doesn't hurt to state that you are interested in XXX or YYY—whatever it is that the lab is working on. But it is better not to expose ignorance than say too little. The conversation with the mentor will be an opportunity to impress him/her with your knowledge.

Once the meeting has been set up, don't assume that means you automatically have a rotation set up. After discussing potential projects (and make sure to do some basic reading before you meet), it is legitimate and important to ask how the lab works: what the mentor expects from the student both during the rotation and afterward as a full-time Ph.D. student in the lab. Some mentors will say they expect students to work until 7 pm every day. Others are more hands-off with regard to time in the lab and vacation. There is a lot of variability, but most mentors expect their students to be as committed as they themselves are to the success of the lab. Make sure you are comfortable with however the lab works.

I found a mentor for my first rotation—can I relax now?

Congratulations on the good work—hopefully you have done your research and vetted your mentor! However, this is not time to relax. To ensure that the mentor will want to offer you a *permanent* position as a graduate student in the lab once the rotation is completed, it is now crucial to bear down, work diligently, show that you are interested, responsible, trustworthy, honest, sociable with the other lab members, and capable of great things to come. It is important to realize that your mentor's time is precious. He/she may be juggling dozens of teaching, administrative and other duties, in addition to mentoring the personnel in the lab. When you have time with your mentor, don't waste it. Be interested. Ask questions. Learn from him/her. Be eager to show your data, but respect the mentor's time. I have seen rotation students receive a cell phone call while talking to a mentor (me!), turn their back and walk away to the hall to talk on the phone. As I mentor, I recognize that there can be an urgent call or emergency, but when this quickly became a pattern, it also became quickly apparent to me that if a rotation student that early on didn't respect me or my time, he/she should find another mentor (who perhaps didn't care or was far enough removed not to notice). Mentors know that any perceived negative behavior by a student during a rotation may only be the tip of the iceberg, and may be very wary about accepting such a student to the lab. The same point is true for students who are deciding whether a lab is a good fit during the rotation—what you see is the tip of the iceberg. If the mentor is stressed has no time for the student during the rotation, he/she will be unlikely to have more time once the rotation is completed.

How else can you impress a mentor? I don't suggest staying until midnight every night, but every mentor knows that most students who (on a day with no classes) work from 9 am to 4 pm will generally get less done than students who come in at 8 am and stay until 6 pm. Be aware that not only the mentor, but the other graduate students, post-docs and even technician in the lab are likely clocking your habits and reporting to the mentor their impressions of you. If you want the job, show that you do! Remember, there may be other students rotating, and perhaps only a single slot open for a PhD position. If you demonstrate that you are the best candidate, you will likely be offered the position.

Final points:

*   Remember, a good healthy environment with a mentor who looks after his/her students is crucial. Even if the field isn't a perfect match with your initial interests, think carefully before turning down such an opportunity for a lab with what appears to be a more interesting project, but with less mentorship.

*   Don't be discouraged if a specific rotation does not work out as well as intended. That's what rotations are for: it's almost like dating, and sometimes it's just not a good match.

*   Do your research before approaching mentors for potential rotations; avoid wasting time in labs that are scientific dead ends. Remember, a mentor who is not particularly committed to mentoring can often appear prolific through collaborations—and in some cases even have significant funding and personnel. But as a student, what is important is not the overall number of papers published with the mentor's name on them, but rather the *PPP—productivity per person* in the lab. A mentor may only publish 2-3 papers a year, but if his/her lab is made up of 2 students and a technician, that is outstanding PPP. Another lab might have 18 people and publish 5-6 papers a year. For this reason it is important for each student to do the research in advance.

*   Finally, remember that you chose this career to enjoy the work/research. If a mentor or lab environment causes anxiety and stress from Day 1, this is probably not an ideal environment for you and I would recommend looking at a different lab.

b.)   On the pathway to success as a graduate student

One of the reasons that I decided to write this manual, was that in my conversations with undergraduate students, I realized that not all of them had a firm understanding of what graduate school is really about. This manual is designed to serve three main purposes: 1) To give an overview of the purpose of graduate school and the rationale for attending it, 2) To provide information that will help undergraduates prepare for the admission process, and 3) To help graduate students understand the timeline and milestones in the course of their graduate studies. I will now address this third point.

Overview of graduate school timeline in a nutshell:

Most graduate programs include a fairly robust load of course-work over the first year or two. Unlike undergraduate studies, however, graduate students take many fewer courses, and are not 'full-time students.' Programs vary, but in first year perhaps (on average) 1-2 hours of classes a day can be a typical load. Some courses are "program-specific" whereas others may be required for all biomedical graduate students. By second year of graduate studies, the courses tend to become more specialized and specific to a graduate program or select field of interest, and in some graduate programs the course–work is completed by the end of the first semester in the second year.

Once courses are completed, usually by the middle or end of the second year, students are often required to take their "Qualifying" or "Comprehensive" examinations. The composition and requirements of these exams vary wildly not just between institutions, but often even between graduate programs within the same institution. The stated goal of the exam is frequently to provide the student with an opportunity to be conceptually challenged by writing a grant application, thus addressing knowledge, critical thinking, reading and writing skills, along with organizational and other abilities that are required for an independent research career. As noted earlier, some programs treat this primarily as an exercise and training opportunity, while others may also view this as an evaluation tool to determine whether a given student has the necessary attributes to continue in the graduate program. The exam is usually administered by a committee selected for each student, and in addition to the research proposal submitted by the student, usually entails an oral exam or defense of the proposal. Depending on how the exam is set-up, the process of completing it can take anywhere from 3-8 months.

Having successfully completed the Qualifying/Comprehensive exam, a student becomes a 'candidate' for the Ph.D., and can now devote all of his/her time to research in the lab, ideally for the next 2-3 years. Toward the end of this period, subject to the program's or institution's rules and requirements for publishing prior to completion of the Ph.D., the student and mentor can decide on a time-frame for writing and submission of the Ph.D. doctoral dissertation—subject to approval by the student's supervisory committee. The usual order of events is that the student submits (often piece by piece) his/her dissertation draft to the mentor for editing and approval, and once the mentor is satisfied, the entire dissertation is submitted to the supervisory committee. Once the committee has edited and/or approved it, the student receives authorization to defend his/her dissertation. This entails a final public seminar on the research accomplished, which is usually followed by a closed-door defense before the supervisory committee. If the committee (often chaired by the mentor) is satisfied, then the student is awarded a doctoral degree. Despite the student's tension and anxiety over the defense, remember that it is very rare for a student to receive authorization from the mentor and supervisory committee to write and defend the dissertation and not be awarded the Ph.D. after the defense.

The first couple years:

Grades:

Undergraduate students are consumed with grades—and rightfully so! We have discussed earlier the significance of obtaining good grades in undergraduate studies to gain access to graduate programs, and undoubtedly this is one of the chief ways in which graduate admissions committees assess potential graduate students. However, once you have been accepted to graduate school and begin a program, you will find that grades, while important, areno longer the only game in town.

In the previous section on rotations, I discussed everything you need to know about selecting and securing a Ph.D. mentor. In the overview above, I discussed the timeline for your Ph.D. What I did not (yet) emphasize is that the processes of researching mentors, finding rotations and doing the actual rotations are all carried out concurrently with your first-year course work. What this means is that if you have a class from 3-4 pm, while you may be tempted to head home for the day to get started on reviewing the material for class, you cannot neglect your lab rotation—if you want to impress the mentor that you are a good fit for the lab. At the same time, as important as it is to show your motivation in the lab, you must be able to pass all your course work. Most programs require a minimal 3.0/B average. Some programs have mostly pass/fail courses, where participation and discussion (perhaps submitting assignments and presenting seminars as well) are the basis for grades, but other programs may still have formal written exams for some courses. In any case, a Ph.D. student must learn to be able to efficiently and effectively do more than walk and chew gum at the same time. While this sounds relatively simple, for some students this can be a difficult balancing act. How much time should you spend reviewing for your course work? Can you afford to spend longer hours in the lab? Navigating these waters can be tricky, and you need to remember the following point: the course work will keep you in graduate school, but it's the research that will define your success propel you out to your next career milestone.

What happens if you do not obtain a B or 3.0 average overall? Different institutions may handle things differently. Typically, if you receive a B- in one course, and you have another course of an equal number of credits with a B+ or better, this will "cancel out" the B- and keep you in good standing. If you receive a grade lower than B-, you may be required to repeat the entire course and obtain at least a B grade. Many institutions put a graduate student on academic 'probation' until the grade has been rectified, and any additional failing grade would be cause for removal from the program. One important thing to note is that aside from you having to retake a course (which takes time from work in the lab), your mentor may lose some respect for you as a student. Since you depend on your mentor for so many things, including recommendation letters that praise your abilities and success as a graduate student, you will want to avoid anything that reduces the mentor's impression of you.

Qualifying/Comprehensive Exam:

Aside from being an exercise and milestone that you must pass in the course of your graduate studies, the Qualifying Exam presents the student with certain opportunities. Many students work intensely in their own labs, and have little external contact with other labs and mentors. However, networking and impressing other faculty is an important goal for ambitious students, who will want to have strong recommendation letters coming from sources other than the mentor. Yet, sometimes there may appear to be limited opportunities to discuss science and make a positive impression with other faculty. The Qualifying Exam is clearly an opportunity to rectify this; the constant contact in the course of those 6-8 months with the examination committee provides motivated students to display their knowledge and critical thinking to other faculty and take advantage of the interaction with the committee.

One thing students need to remember; during the course of this long exam, research goes on, and the student is expected to continue driving his/her research project forward. Some mentors will be more accommodating than others, but students who expect that they can significantly cut down experimental work in the lab to work on the

Qualifying Exams may be disappointed to find an unhappy mentor. Mentors, who are responsible for evaluating graduate students and writing recommendation letters, will maintain that graduate students need to learn to juggle different tasks and manage their time efficiently. As an example, when I was a student, I made a point of never discussing course-work, Qualifying Exams and other student-obligations with my mentor, but rather focused exclusively on discussing our common research with her. My goal was to show her that I am professional, and that I am capable of carrying out those student requirements without slowing down my research—with the knowledge that one day she would be called to write recommendation letters. Yes, that type of thinking might be exaggerated, but it provides an example of how a student can impress faculty members by being highly professional.

Fellowships:

In the course of graduate studies, many students have an opportunity to draft a fellowship application, often known as "pre-doctoral fellowships." Some of the fellowships may be awarded by the institution (if there is funding committed to such an endeavor). Other fellowships may be obtained from the National Institutes of Health (for example F31 fellowships) or other funding foundations, such as the American Cancer Society or American Heart Association. Obviously, obtaining an 'external fellowship' from outside your institution is considered more prestigious, but any award recognizing you and your research is very important. Some mentors may push students to apply for external fellowships, because if successful, the mentor will be able to "save" the money paid toward your stipend (it comes from the fellowship), thus freeing up grant and other money for supplies, equipment (and potentially even another student). However, other mentors may be flush with grant money, and the time and energy involved in editing your proposal and filling out the myriad list of recommendations and supporting information may not seem worthwhile. It is up to you as the student to take charge and organize the submission (with the mentor's approval and support) if you want it. Note: many fellowship opportunities, including those F31 fellowships from the NIH, are exclusively available for US citizens or those with permanent residence. In addition, some institutes at the NIH, such as

General Medical Sciences, do not offer F31 fellowships for most pre-doctoral students, choosing instead to put the bulk of support into funding training grants for graduate programs.

Finishing up and dissertation defense:

Most students are anxious to complete their training as a Ph.D. student for a variety of reasons.

These include the satisfaction of completing their degree, the ability to move on with one's life, the opportunity to get a more permanent and better paying position, and potentially, a chance to

leave behind (what in some cases might be) a frustrating situation. However, it is critical for students to remember their long-term goals in the completion of a Ph.D. degree before hurrying

across the finish line.

By long-term goal, my meaning is that the degree itself is not the long-sought prize, but rather what the degree symbolizes. If the degree comes with great training, strong publications and positive recommendation—terrific. However, if the completion of one's degree comes without publications, for example, it becomes a

Pyrrhic degree. In other words, as a student, you need to consider whether you should be in a great hurry just to receive your diploma. Students in their advanced years as Ph.D. students are typically far more efficient in research. The combination of experience and wisdom gained, coupled with completion of courses and more time for research often make the last few years the most productive ones for a student. If an extra 6-8 months might a difference and provide an opportunity to complete/submit an additional paper or papers, the possibility of staying on for up to a year might be worthwhile. Another concern is whether the student has lined up a job (post-doctoral position or other job)—most scientists would not recommend completion of the Ph.D. before a job had been lined up. In any case, the best scenario is to discuss timelines with your mentor long before you plan to graduate (a year or 18 months in advance is a good guideline). While the timeline can change depending on the situation in the lab, having an open and ongoing conversation about these important and sensitive issues with your mentor, well in advance, is highly advised.

6.) CONCLUSIONS

Hopefully this manual will serve as a resource for undergraduate students and provide some necessary information for students to make an informed decision about whether and how to apply for graduate programs. In addition, the manual should also help students to understand the various steps along the application process, as well as to guide students who have been accepted into a graduate program. Key to success at all levels is the student's ability to communicate, ask questions and successfully network and navigate through these uncharted waters. In today's competitive environments, even the best and brightest students will need to be cognizant of the need for outstanding communications with advisors, mentors and others, to ensure maximal success in their careers.

www.ingramcontent.com/pod-product-compliance
Lightning Source LLC
Chambersburg PA
CBHW031529210526
45463CB00011B/2384